Scientific Progress

Miguel A. Sanchez-Rey

The Perils of New Atheism:

Ideological War-Crime and Propaganda

By

Miguel A. Sanchez-Rey

There is never the belligerence of New Atheism's advocacy of radical activism. Radical activism which promotes a militant approach to advocating atheist thought. But radical activism is an ideological motivation. An ideological motivation that furthers a violent confrontation towards religious beliefs and superstitious ideals.

New Atheism started with three key players: Richard Dawkins, Daniel Dennett, and the late Christopher Hitchens, as a counter-reaction to the growing animosity toward secular policy and rationalist beliefs. Valid in its core themes of liberal progressivism they nevertheless shared ideological similarities with much of the conservative militant establishment. Favoring policies of domination and neo-colonial imperialism which stresses an aggressive approach that fights back at religious beliefs and democratic socialism. An affinity to state-capitalist propaganda and the war-machine. An attachment to the religious state and an adherence to state policies that favor free-markets. Others like Rebecca Newberger Goldstein, Steven Pinker, Sam Harris, and Massimo Pigliucci, also subscribe to these core political and social views, though without qualifications or unanimity.

A cult following ensued that brought world-wide fame to the characters of New Atheism. A following generated by the economic crisis of summer of 2011. Richard Dawkins and Christopher Hitchens assertion that the rise of radical Islamic fundamentalism should be seen as a serious threat to world-peace, and as the root cause of the geo-political instability in the Middle East, is ideology. Their base, ever more incited by their promotion of violent activism, soon became a toxic mixture of radicalist and extremist with elements of neo-Fascism

2

that within four years, up to the Brexit of 2016, stirred the white supremacist in Western Europe that opposed the internationalist model of the European Block and the plight of refugees that immigrated from the Middle East to Western Europe. And only then encouraged the American political establishment, including the American propaganda machine, i.e. NY Times, Wall Street Journal, Foreign Affairs, etc., to side with New Atheism's false social policies. In which by false social policies is meant false inclinations that are true but not motivationally true.

Perils which are consequential and in which there seems to be no end to the immediate ineffectiveness of the militant approach. Their promotion of ideological war-crime, that in which they hold no respectability to the internationalist consensus of the diplomatic approach and in which there is no regard to international law dictating the need for human rights, has backfired on the Militant Atheists as fundamentalism, in the form of political and religious fundamentalism, has adapted to New Atheism's social and war-like policies of aggression and violent opposition. In doing so the key players have only cause more serious harm than good and has unleash global economic and political instability that has encouraged the solidification of authoritarianism. Posing a severe existential threat to social democracy in the form of a benign propaganda machine that further exacerbates hate groups that are disinclined to the internationalist model.

The Depth and Dynamics of the Sociopathic Norm

[Author: Miguel A. Sanchez-Rey]

The sociopathic norm is a world-wide crime ring of humanitarians and academics involve in both financial and academic fraud. They went world-wide at the dawn of the Scientific Age with the advent of the Windows XP high-tech revolution. It wasn't clear at the time, which manifested as an organize criminal element, the depth of this cult but with certainty many unqualified and fringe [or groups of] academics and humanitarians establish themselves into positions of decision-making as leading authorities in their fields.

But at the time what was thought of as relaxing the requirements in the employment of academics and financial experts, that would open doors of opportunity for upward mobility and economic growth, backfired at a world-wide scale. More unqualified experts and academics made bad decision-making that has had a long-term impact to global national security and planetary tranquility.

A cult that can be hard to spot but when spotted can be frighteningly messianic. Exhibiting patterns of behavior that is unprecedented and consistent with a cult of the Scientific Age. Those patterns are (1) clueless clues, (2) ingrained behavior, (3) preying upon, (4) desire to control others and each other, (5) lack of conscience, (6) narcissism, (7) paperback writers and plagiarists, (8) sycophants and cronies, (9) internet addicts, (10) average intelligence in relation to academic and financial occupation but nevertheless memorizers that become better at recalling and restating information as they get older, (11) fraudulent life-styles, (12) capable of genocide when giving the opportunity and (13) manic in disposition.

The cult can be spotted when there is suspicious behavior amongst academics and humanitarians. In particular, they can be so involved with each other without realization but when realize the sociopathic norm becomes self-aware as an organized crime ring. In such a way that they begin to prey and brutally control each other. Since their behavior is ingrained at an early stage the sociopathic norm is oblivious to their own capability of committing horrid acts of hatred and malice. While also being silly and benign in nature with false claims of benevolence and altruism.

Internet addicts they utilized social networking sites to further their reputations. Claiming expertise, they provide internet commentary on various subjects while incapable to excel in most of the varied subjects they are presenting outside their own academic and financial specialization. They are also too quick to publish by using fraudulent methods to accelerate their writing.

Rooting out this cult at a world-wide scale requires increasing standards in academia and finance. Most importantly when the cult is spotted it is imperative that they be broken apart in such a way that their assets be confiscated, their titles be annulled and then face criminal prosecution so that they pose no further rabid threat to the Scientific Age. Preventing mass-suicide amongst their own followers.

Understanding the depth of sociopathic norm is at an early stage but with further studies on the dynamics of this cult will, in the long-term, neutralize them from causing any further harm to planetary tranquility and the Scientific Age. Short-term wise restoring genuine social harmony.

Metamorphic Topological Schemes

Miguel A. Sanchez-Rey

Abstract

Establish topological schemes in metamorphic space as A-scheme and B-scheme.

January 30th, 2017

Topological Scheme in Metamorphic Space

     Metamorphic space is define as cosmological homotopic states between variant [stringy]'s of perfect number. One can relate each variant of perfect number using the procedure of A-scheme and B-scheme. Where the N [n] of A-scheme and B-scheme is of prime. Such that A-scheme is variant [of stringy]'s containing color black and B-scheme is variant [of stringy]'s contains color white. Interelating and switching variant [of stringy]'s of prime using A-scheme and B-scheme, in metamorphic space, ignites the terraformic process in which by prime factorization the variant [of stringy]'s do not endlessly replicate in The Grand Unification Scheme. By utilizing SUPREME one imposes a homogeneous topology between the base space and target space; and vice versa. Thereby relating A-scheme and B-scheme one achieves a topological scheme in metamorphic space.

Star Trek is War-Like

[Author: Miguel A. Sanchez-Rey]

Star Trek is a science-fiction genre set in the distant future, around the 22nd to 24th century, that takes astronauts on an exploration mission across many planets that addresses problems that are metamorphic to the historical consciousness of the mid-20th and early early-21th century. Utilizing science-related themes with highly speculative technological ideas, that help to carry out the many stories of the Star Trek universe, would-be astronauts enlist in Star Fleet under the leadership of the United Federation of Planets headquarter on planet Earth.

The United Federation of Planets was founded to unite the known planets, capable of warp-drive, that came into first contact in earlier era of the star trek universe. That by the prime directive only planets capable of warp-drive are qualified to make first contact with Star Fleet.

Conflict in Star Trek subsume the majority of the plots in the star trek universe. Conflict with alien intelligences that ravishes the star trek universe. For example: the Klingons relationship with the United Federation of Planets has been relatively unstable up to the present and the Borg has been an existential threat to the Star Trek universe without end.

The star trek genre thrives in the idea of conflict. While it also addresses social and technological problems without embedding conflict into the star trek universe there is no entertainment value to the star trek universe. As to motivate the star trek universe requires the idea that war is good and peace is bad. War is good because it sustains a war-like society that thrives in conflict. Peace is bad because no purpose is to be gain from a highly-technological sedentary society. The interplay between conflict and self-actualization is at the heart of the Star Trek universe. But one can be misled to believe that what appears to be an innocent interplay between self-actualization and conflict, as James T. Kirk's self-realization as

2

the captain of the U.S.S. Enterprise or Data's desire to be more human, in the Star Trek

Generation genre, is in actual a prelude to war-crime. As to achieve self-actualization and to

reach the climax of the star trek story war-crime must be committed on a large-scale to prove a

point. A point that without conflict; without pain and grief, there can be no self-actualization

and yet to self-actualize in star trek requires, in most cases, a large-scale conflict that unleashes

havoc to planet Earth and all other alien worlds. That to reach the peak of self-actualization the

star trek universe must undergo the grief of the death of countless many that number at a

death-toll proportional to that of a giant Earth-like planet.

Even then star trek's desire for exploration has skewed the boundaries that limit

encroaching into other solar habitats or other alien occupied regional systems. Encroaching

into alien worlds sets the stage for a massive conflict and propels the story of the star trek

universe. As the captain of the enterprise must constantly make two crucial choices: avoid

conflict or take action? Obey Star-Fleet command or disobey? That is whether or not the

captain of Enterprise is willing to be implicated in war-crime and face the dire consequences

that amount to nothing, no less and no more, than the false realization of self-actualization in

which in truth the captain should be brought up on charges and face harsh prosecution in the

Federation Council.

The unstable nature of the star trek universe makes the plot and stories of star trek

suspect. Even though it satisfies one's desire to see the transitional future it nevertheless is a

genre of metaphor that relates to present-day historical trauma's and ordeals. Though it

reveals ideas with strong liberal themes it nevertheless mixes them with the concept of the

antagonist and protagonist element of the modern film plot. That is to say the one is not

3

dealing with an innocent genre of would-be astronauts that strive to self-actualize and reach for the stars by addressing modern social and political themes but the willingness to take part in conflict, on both a local and galactic level, to achieve a sense of fulfillment while ignoring and skewing the reality of the death of hundreds, thousands and millions that would inevitably land the captain of the enterprise in a war-crime tribunal either on planet Earth or in an alien planet facing life-in-prison or execution for ignoring orders from Star Fleet command that could potentially have save the lives of countless millions for the good of the few.

It is nevertheless that the star trek universe is sadist in its core nature and propelling the sadist element of the star trek universe misled it's many fans to embrace star trek's diverse stories of conflict and false sense of self-actualization. In truth one is to see star trek in light of the reality of conflict. That conflict is not to be desired or embraced but to be avoided at all cost. That war is not to be encouraged and glorified. A war-like society, as the star trek universe, is a society that is constantly at war and strives to encourage and instigate conflict in order to achieve further societal development and fulfillment. Within only a matter of months a society that is constantly at war will be, in all likely, wiped out. And for those reasons the longevity of the star trek universe is short-live.

The Masters of Last Resort

By

Miguel A. Sanchez-Rey

Military action is the last resort to a large-scale conflict between nation-states and to protect the state from an internal threat that could jeopardize the ruling power.

Monarchy is an earlier response to the state-of-nature. In which the ruling regime is giving dominion over a populace base on lineage and ownership of property. Military action is to be decided by a monarchy when an external threat is apparent. It is to be dispatch when an internal strife becomes a threat to the ruling power. And it is to protect the ruling class at all cost to secure the lineage of the state.

The ruling regime lineage spans more than two thousand years since the birth of Christ. With the Egyptian pharaohs, the Emperors of the Roman Empire and the Holy Roman Empire, the monarchs of early classical Europe, and at present the remaining monarchs of the Middle East and the East Asia peninsula.

The beginning of the Enlightenment brought forth the realization of parliamentary and constitutional democracy. That in which the populace achieves control of the state and in which the affairs of state are to be carried out by elected politicians on their behalf. That in which the population has a say in the future of democracy and in which the rights of future generations override the lineage of the monarchy.

The tyranny of monarchy has been well-known. Monarchs take brutal action to protect their lineage and dominion. Law and order, in a state ruled by a monarchy, is swift to control their population so that the interest of the ruling regime is carried out and protected. The

ruling regime of a monarchy has first say on the affairs of security of the state and in which

treaties are agreed upon.

The modern era has seen the decolonization of states and has witnessed the transition

of monarchy into parliamentary and constitutional democracy on a world-wide scale. Royalty

has become a connotation of duty to the state in return for privilege.

Monarchy has been well-known to relate to the bourgeoisie. And in which the

proletariat overthrows the bourgeoisie in order to control the means of production and the

affairs of state. That in which the protection of the state is carried out by the populace and in

which the ruling parliamentary and constitutional government decides to take military action to

protect the interest of the ruling government and tranquility of the governed.

In that manner monarchs become the heads-of-state of parliamentary government.

They reside as dignitaries of the state. They take action only upon authorization of the

democratic regime. They are nevertheless a last resort to secure the tranquility of the state but

nevertheless the ideal. For monarch's rule by force and brutality to secure their lineage but as

well must protect their dominion.

The control of the use of force is now the purview of the democratic state. For in which

the democratic state protects itself from an external foreign threat by utilizing military force.

They may take excessive force or resort to unilateral action to carry out foreign interests. They may invade countries upon majority as well.

Whereby the heads of state of a parliamentary democratic government do not interfere in the affairs of state including in the foreign interests of the ruling regime. But where two ruling regimes exist a coup takes over and the democratic state is confronted with the brutal action of the heads of state. Thereby the populace becomes no longer citizens of the state but subjects to the heads of state. But where the heads of state are the ruling regime the democratic state becomes an authoritarian society.

In which the dictatorial regime imposes itself on the population by force and in which the last resort is to be carried out at their discretion. A revolt ensues and the state is met with existential futility. It's then that the last resort of the state is to be enacted by the democratic government. In which the heads of state represent the interest of the state. Whereby the heads of state represent, as well, military forces on behalf of the interests of the democratic regime.

Military action is always the last resort to an imminent or impending conflict. That is the ruling government gives the authorization to take military action. The heads of state cooperate to do exactly what the state demands and in terms of military interests they are to carry out the foreign interests of the democratic state but are not to interfere in the foreign affairs of the ruling party.

The right to last resort is no longer the purview of the heads of state but the dominion of the democratic regime. Whereby the heads of state are no longer masters of the state but representatives of the state. And in which rulership belongs to the elected government and in which they alone decide foreign policy. That in which military forces are subjects to the scientific state and in which the heads of state are no longer the arbitrators of last resort. But in which the last resort is to decided by the democratic regime and in which they alone hold mastership over the heads of state and bring about the protection of the state from internal and foreign strife.

The First Task of PHPR

From String Theory to Topological Strings in Metaspace

Miguel A. Sanchez-Rey

The First Task Task of PHPR [The Physicalist Program] is a 100 year task. It's a task that is a collaborative effort amongst high-energy physicist to resolve the long-standing issue of mineral depletion that also includes an international effort to gain access to metaspace. In conjunction it includes utilizing the International Thermonuclear Experimental Reactor [ITER] as a 40 year window of opportunity in which sixty percent of The First Task must be completed.

Not only does it involve the completion of The Grand Unification Scheme (which is the second most important task) but also gaining access to metaspace that will result in igniting the terraformic process. It's a stringent task but a task well worth pursuing as the pay-offs are limitless in extent. Not only will it resolve mineral depletion but it will also accelerate advances in the technological and engineering sciences.

There can be no anticipation of metaspace until one gains access to metaspace however efforts are to be put into place that includes computational control and SUPREME, prime factorization, a guide through metaspace, and the A-scheme and B-scheme. Metaprocedures that allows one to control and harness metaspace.

But metaspace is defined as cosmological homotopic states between variant [of stringy]'s of prime. These variant [of stringy]'s are understood to be variations of supersymmetric string. Where one can say that SUSY-like physics resides in metaspace. In such a way that the first stringy is a membrane solution of all other three string models. Where each string model is a first quantized solution that includes open and close strings in a 11-dimensional membrane that can be related by duality transformations (or mirror symmetry). A unique transformation led to the discovery of the first variant [of stringy] in the form of a Chern-Simons solution in N = 4 Super-Yang Mills.

At first it was seen as a parameterized solution to a holographic counterterm. With further advances it was then discovered that this crude solution is a variant and that by perturbing this variant all but 32 variants remain unaccounted for.

String theory is the dominant field of research in the string landscape. But string theories drawbacks are tantamount. Even as it makes significant cross-roads in supersymmetry and quantum cosmology one has yet to observe or manipulate a one-dimensional string residing on a world-sheet manifold. Yet strings includes a monumental amount of depth on a mathematical level. Depth that has serve its purpose. Giving one significant access to pure physics that ultimately led to metaspace. Even then string theory will prove worthy of study for making experimental predictions not only in the 11-dimensional membrane universe but also at 3+1 low-energy classical scale.

Leading to abandoning string theory in favor of string topology in hopes of furthering advances in metaspace. String topology, unlike string theory, develop by Edward Witten and Cumrum Vafa, includes monumental eloquence and simplicity that can achieve faster results in gaining knowledge about the mechanisms of all the known variant [of topstringy]'s while by applications of all the metaprocedures allows one to control metaspace once access to metaspace is achieved at the higher-energy scale.

It should be stated that variant [of topstringy]'s is subject to computational control and SUPREME. That implies that mathematical understanding of string topology will change and it will not be clear until one gains access to metaspace what string topology may entail as variant [of topstringy] but theoretical advances in topstringy will be essential to controlling metaspace in which not being able to anticipate metaspace hinders any hindsight of these variants. Only by using what is already recognized as a topological theory of strings, as the A-model and B-model, in the gravitational supersymmetric Kähler and Calabi-Yau manifold, will continuing applications of string topology to The First Task of PHPR yield limitless positive results in metaspace.

Definition of the Grand Unification Scheme:

The collection of variant [of topstringy]' s of prime that exist in $\emptyset \subseteq$ H_m that can related and interchanged by their topological solitons in the D − variant manifold.

www.ingramcontent.com/pod-product-compliance
Lightning Source LLC
Chambersburg PA
CBHW081820170526
45167CB00008B/3480